MAN AND TECHNICS

A Contribution to a Philosophy of Life

Oswald Spengler

Albatross Publishers
Naples, Italy
2020

ISBN 978-1-946963-48-2

Copyright 1932 by Alfred A. Knopf, Inc.

Originally published as
DER MENSCH UND DIE TECHNIK
Copyright by C. H. Beck'sche Verlagsbuchhandlung
Munich, 1931

PREFACE

IN the following pages I lay before the reader a few thoughts that are taken from a larger work on which I have been engaged for years. It had been my intention to use the same method which in *The Decline of the West* I had limited to the group of the higher Cultures, for the investigation of their historical pre-requisite — namely, *the history of Man from his origins*. But experience with the earlier work showed that the majority of readers are not in a position to maintain a general view over the mass of ideas as a whole, and so lose themselves in the detail of this or that domain which is familiar to them, seeing the rest either obliquely or not at all. In consequence they obtain an incorrect picture, both of what I have written and of the subject-matter about which I wrote. Now, as then, it is my conviction that the destiny of Man can only be understood by dealing with all the provinces of his activity *simultaneously and comparatively*, and avoiding the mistake of trying to elucidate some problem, say, of his politics or his religion or his art, solely in terms of particular *sides* of his being, in the belief that, this done, there is no more to be said. Nevertheless, in this book I venture to put forward some of the questions. They are a few among many. But they are interconnected, and for that reason may serve, for the time being, to help the reader to a provisional glimpse into the great secret of Man's destiny.

CONTENTS

PREFACE

I. TECHNICS AS THE TACTICS OF LIVING   3
Process and means. The contest and the weapon. Evolution and fulfilment. Passingness as the form of the actual.

II. HERBIVORES AND BEASTS OF PREY   19
Man as beast of prey. The spoil and the spoiler. Movement, as flight and onset. The predatory eye and its world. Unalterable genus-technique and inventive human technique.

III. THE ORIGIN OF MAN: HAND AND TOOL   35
The hand as organ of touch and of deed. Differentiation of the making and the using of the weapon. Liberation from the compulsion of the genus. "Thought of the eye" and "thought of the hand." Means and aim. Man as creator. The single act. Nature and "art." Human technique artificial. Man *versus* Nature. The Tragedy of Man.

IV. THE SECOND STAGE: SPEECH AND ENTERPRISE   49

>Collective doing. How old is speaking in words? Purpose of speech, collective enterprise. Purpose of enterprise, the enhancement of human power. Separation of thought and hand, leader's work and executant's work. Heads and hands, the hierarchy of talents. Organization. Organized existence, state and people, politics and economics. Technics and human numbers. Personality and Mass.

V. THE LAST ACT: RISE AND END OF THE MACHINE CULTURE   75

>Vikings of the intellect. Experiment, working hypothesis, perpetual motion. Meaning of the machine, the inorganic forces of Nature compelled to work. Industry, wealth, and power. Coal and population. Mechanization of the world. Symptoms of the decline, diminution of leader-natures. Mutiny of the hands. The lost monopoly of technics. The coloured world. The End.

Chapter I

※

TECHNICS AS THE TACTICS OF LIVING

ONE

THE problem of technics and its relation to Culture and to History presents itself for the first time only in the nineteenth century. The eighteenth, with its fundamental scepticism — that doubt that was wellnigh despair — had posed the question of the meaning and value of *Culture*. It was a question that led it to ever wider and more disruptive questions and so created the possibility for the twentieth, for our own day, of looking upon the *entirety* of world-history as a problem.

The eighteenth century, the age of Robinson Crusoe and of Jean Jacques Rousseau, of the English park and of pastoral poetry, had regarded " original " man himself as a sort of lamb of the pastures, a peaceful and virtuous creature until Culture came to ruin him. The technical side of him was completely overlooked, or, if seen at all, considered unworthy of the moralist's notice.

But after Napoleon the machine-technics of Western Europe grew gigantic and, with its manufacturing towns, its railways, its steamships, it has forced us in the end to face the problem squarely and seriously. What is the significance of technics? What meaning within history, what value within life, does it possess, where — socially and metaphysically — does it stand? There were many answers offered to these questions, but at bottom these were reducible to two.

On the one side there were the idealists and ideologues, the belated stragglers of the humanistic Classicism of Goethe's age, who regarded things technical and matters economic as standing outside, or rather *beneath*, " Culture." Goethe himself, with his grand sense of actuality, had in *Faust II* sought to probe this new fact-world to its deepest depths. But even in Wilhelm von Humboldt we have the beginnings of that anti-realist, philological outlook upon history which in the limit reckons the values of a historical epoch in terms of the number of the pictures and books that it produced. A ruler was regarded as a significant figure only in so far as he passed muster as a patron of learning and the arts — what he was in other respects did not count. The State was a continual handicap upon the true Culture that was pursued in lecture-rooms, scholars' dens and studios. War was scarcely believed in, being but a relic of bygone barbarism, while economics was something prosaic and stupid and beneath notice, although in fact it was in daily demand. To mention a great merchant or a great engineer in the same breath with poets and thinkers was almost an act of *lèse-majesté* to "true" Culture. Consider, for instance, Jakob Burckhardt's *Weltgeschichtliche Betrachtungen* — the outlook is typical of that of most professors of philosophy (and not a few historians, for that matter), just as it is the outlook of those literates and æsthetes of today who view the making of a novel as something more important than the designing of an aircraft-engine.

On the other side there was Materialism — in its essence an English product — which was the fashion among the half-educated during the latter half of the nineteenth century, and the philosophy of liberal journalism and radical mass-meetings, of Marxist and social-ethical writers who looked upon themselves as thinkers and seers.

If the characteristic of the first class was a lack of the sense of reality, that of the second was a devastating shallowness. Its ideal was *utility*, and utility only. Whatever was useful to " humanity " was a legitimate element of Culture, *was* in fact Culture. The rest was luxury, superstition, or barbarism.

Now, this utility was utility conducive to the " happiness of the greatest number," and this happiness consisted in not-doing — for such, in the last analysis, is the doctrine of Bentham, Spencer, and Mill. The aim of mankind was held to consist in relieving the individual of as much of the work as possible and putting the burden on the machine. Freedom from the "misery of wage-slavery," equality in amusements and comforts and " enjoyment of art " — it is the *panem et circenses* of the giant city of the Late periods that is presenting itself. The progress-philistine waxed lyrical over every knob that set an apparatus in motion for the — supposed — sparing of human labour. In place of the honest religion of earlier times there was a shallow enthusiasm for the " achievements of humanity," by which nothing more was meant than progress in the technics of labour-saving and amusement-making. Of the soul, not one word.

Now, such ideals are not at all to the taste of the great discoverers themselves (with few exceptions), not even to that of the finished connoisseurs of technics. It is that of the *spectators* around them who, themselves incapable of discovering anything (or anyhow of understanding it if they did), sense that there is something to their own advantage in the wind. And out of these conditions, since in every " civilization "[1] materialism is distinguished by its lack of imaginative power, there is formed a picture of the future in which the ultimate object and the final permanent condition of humanity is an Earthly Paradise conceived in terms of the technical vogue of, say, the eighties of last century — a rather startling negation, by the way, of the very concept of progress, which by hypothesis excludes " states." This order of ideas is represented by books like Strauss's *Alte und Neue Glaube,* Bellamy's *Looking Backward,* and Bebel's *Die Frau und der Sozialismus.* No more war; no more distinctions of law, peoples, states, or religions; no criminals or adventurers; no conflicts arising out of superiorities and unlikenesses, no more hate or vengeance, but just unending comfort through all millennia. Even today, when we are still living out the last phases of this trivial optimism, these imbecilities make one shudder, thinking of the

---

[1] The word is used, of course, in the specific sense which it bears throughout *The Decline of the West.—Tr.*

appalling boredom — the *tædium vitæ* of the Roman Imperial age — that spreads over the soul in the mere reading of such idylls, of which even a partial actualization in real life could only lead to wholesale murder and suicide.

Today both views are out of date. At last, with the twentieth, we have come to a century that is ripe enough to penetrate the final significance of the *facts* of which the totality constitutes *world-history*. Interpretation of things and events is no longer a matter of the private tastes of individuals of a rationalizing tendency, or of the hopes and desires of the masses. The place of " it shall be so " and " it ought to be so " is taken by the inexorable " it is so," " it will be so." A proud skepsis displaces the sentimentalities of last century. We have learned that history is something that takes no notice whatever of our expectations.

It is physiognomic tact, as I have elsewhere called it [1] — namely the quality which alone enables us to probe the meaning of all happening — the insight of Goethe and of every born connoisseur of men and life and history throughout the ages — that reveals in the individual his or its deeper significance.

---

[1] *The Decline of the West*, English edition, Vol. I, p. 100.

TWO

IF we are to understand the essence of Technics, we must not start from the technics of the machine age, and still less from the misleading notion that the fashioning of machines and tools is the *aim* of technics.

For, in reality, technics is immemorially old, and moreover it is not something historically specific, but something immensely general. It extends far beyond mankind, back into the life of the animals, indeed of *all* animals. It is distinctive of the animal, in contrast to the plantwise, type of living that it is capable of moving freely in space and possesses some measure, great or small, of self-will and independence of Nature as a whole, and that, in possessing these, it is obliged to maintain itself against Nature and to give its own being some sort of a significance, some sort of a content, and some sort of a superiority. If, then, we would attach a significance to technics, we must start from the *soul*, and that alone.

For the free-moving life of the animal[1] is struggle, and nothing but struggle, and it is the *tactics* of its living, its superiority or inferiority in face of "the other" (whether that "other" be animate or inanimate Nature), which decides the *history* of this life, which settles whether its fate is to suffer the history of others or to be itself their history. *Technics is the tactics of living;* it is the inner form of which the *procedure* of conflict — the conflict that is identical with Life itself — is the outward expression.

This is the second error that has to be avoided. *Technics is not to be understood in terms of the implement.* What matters is not how one fashions things, *but what one does with them;* not the weapon, but the battle. Modern warfare, in which the decisive element is tactics — that is, the technique of running the war, the techniques of inventing, producing, and handling the weapons being only items in the process as a whole — points a general truth. There are innumerable techniques in which no implements are used at all, that of a lion outwitting a gazelle, for instance, or that of diplomacy. Or, again, the technics of administration, which con-

---

[1] *Decline of the West,* English edition, Vol. II, p. 3.

sists in keeping the State in form for the struggles of political history. There are chemical and gas-warfare techniques. Every struggle with a problem calls for a logical technique. There is a technique of the painter's brush-strokes, of horsemanship, of navigating an airship. Always it is a matter of *purposive activity*, never of *things*. And it is just this that is so often overlooked in the study of prehistory, in which far too much attention is paid to things in museums and far too little to the innumerable processes that must have been in existence, even though they may have vanished without leaving a trace.

Every machine *serves* some one process and owes its existence to *thought about this process*. All our means of transport have developed out of the *ideas* of driving and rowing, sailing and flying, and not out of any concept such as that of a wagon or of a boat. Methods themselves are weapons. And consequently technics is in no wise a " part " of economics, any more than economics (or, for that matter, war or politics) can claim to be a self-contained " part " of life. They are all just *sides of one active, fighting, and charged life*. Nevertheless, a path does lead from the primeval warring of extinct beasts to the processes of modern inventors and engineers, and likewise there is a path from the trick, oldest of all weapons, to the design of the machines with which today we make war on Nature by outmanœuvring her.

Movement on these paths we call Progress. This was the great catchword of last century. Men saw history before them like a street on which, bravely and ever forward, marched "mankind" — meaning by that term the white races, or more exactly the inhabitants of their great cities, or more exactly still the "educated" amongst them.

But whither? For how long? *And what then?*

It was a little ridiculous, this march on infinity, towards a goal which men did not seriously think about or clearly figure to themselves or, really, *dare* to envisage — *for a goal is an end*. No one does a thing without thinking of the moment when he shall have attained that which he willed. No one starts a war, or a voyage, or even a mere stroll, without thinking of its direction and its *conclusion*. Every truly creative human being knows and dreads the *emptiness* that follows upon the completion of a work.

To development belongs fulfilment — every evolution has a beginning, and every fulfilment is an *end*. To youth belongs age; to arising, passing; to life, death. For the animal, tied in the nature of its thinking to the present, death is known or scented as something in the future, something that does *not* threaten it. It only knows the fear of death *in the moment* of being killed. But man, whose thought is emancipated from the fetters of here and now, yesterday and tomorrow, boldly investigates the " once " of past and future, and it depends on the depth or shallowness of his nature whether he triumphs over this fear of the end or not. An old Greek legend — without which the Iliad could not have been — tells how his mother put before Achilles the choice between a long life or a short life full of deeds and fame, and how he chose the second.

Man was, and is, too shallow and cowardly to endure the fact of the *mortality* of everything living. He wraps it up in rose-coloured progress-optimism, he heaps upon it the flowers of literature, he crawls behind the shelter of ideals so as not to see anything. But impermanence, the birth and the passing, is the *form of all that is actual* — from the stars, whose destiny is for us incalculable, right down to the ephemeral concourses on our planet. The life of the individual — whether this be animal or plant or man — is as perishable as that of peoples of Cultures. Every creation is foredoomed to decay, every thought, every discovery, every deed to oblivion. Here, there, and everywhere we are sensible of grandly fated courses of history that have vanished. Ruins of the "have-been" works of dead Cultures lie all about us. The hybris of Prometheus, who thrust his hand into the heavens in order to make the divine powers subject to man, carries with it his fall. What, then, becomes of the chatter about "undying achievements"?

World-history bears a very different face from that of which even our age permits itself to dream. The history of man, in comparison with that of the plant and animal worlds on this planet — not to mention the lifetimes prevailing in the star world — is brief indeed. It is a steep ascent and fall, covering a few millennia, a period negligible in the history of the earth but, for us who are born with it, full of tragic grandeur and force. And we, human beings of the twentieth century, go downhill *seeing*. Our eye for history, our faculty of writing history, is a revealing sign that our path lies downward. At the peaks of the high Cultures, just as they are passing over into Civilizations, this gift of penetrating recognition comes to them for a moment, and only for a moment.

Intrinsically it is a matter of no importance what is the destiny, among the swarms of the " eternal " stars, of this small planet that pursues its course somewhere in infinite space for a little time; still less important, what moves for a couple of instants upon its surface. But each and every one of us, intrinsically a null, is for an unnamably brief moment a lifetime cast into that whirling universe. And for us therefore this world-in-little, this " world-history," is something of supreme importance. And, what is more, the *destiny* of each of these individuals consists in his being, by birth, not merely brought into this world-history, but brought into it in a particular century, a particular country, a particular people, a particular religion, a particular class. It is *not* within our power to choose whether we would like to be sons of an Egyptian peasant of 3000 B.C., of a Persian king, or of a present-day tramp. This destiny is something to which we have to adapt ourselves. It *dooms* us to certain situations, views, and actions. There are no " men-in-themselves " such as the philosophers talk about, but only men of a time, of a locality, of a race, of a personal cast, who contend in battle with a *given* world and win through or fail, while the universe around them moves slowly on with a godlike unconcern. This battle *is* life — life, indeed, in the Nietzschean sense, a grim, pitiless, no-quarter battle of the Will-to-Power.

Chapter II

HERBIVORES AND BEASTS OF PREY

THREE

MAN *is a beast of prey.* Acute thinkers, like Montaigne and Nietzsche, have always known this. The old fairy-tales and the proverbs of peasant and nomad folk the world over, with their lively cunning: the half-smiling penetration characteristic of the great connoisseur of men, whether statesman or general, merchant or judge, at the maturity of his rich life: the despair of the world-improver who has failed: the invective of the angered priest — in none of these is denial or even concealment of the fact as much as attempted. Only the ceremonious solemnity of idealist philosophers and other . . . theologians has wanted the courage to be open about what in their hearts they knew perfectly well. Ideals are cowardice. Yet, even from the works of these one could cull a pretty anthology of opinions that they have from time to time let slip concerning the beast in man.

Today we must definitely settle accounts with

this view. Scepticism, the last remaining philosophical attitude that is possible to (nay, that is *worthy of*) this age, allows no such evasion of issues. Yet, for this very reason, neither would I leave unchallenged other views that have been developed out of the natural science of last century. Our *anatomical* treatment and classification of the animal world is (as is to be expected from its origin) dominated entirely by the materialist outlook. Granted that the picture of the body, as it presents itself to the human (and only to the human) eye, and *a fortiori* that of the body as dissected and chemically treated and experimentally maltreated, eventuates in a system — the system founded by Linnæus and deepened in its palæological aspect by the Darwinian school — a system of static and optically appreciable details, yet after all there *is* another, a quite other and unsystematic, ordering according to species of *life*, which is revealed only through unsophisticated living with it, through the inwardly felt relationship of Ego and Tu, which is known to every peasant, but also to every true artist and poet. I love to meditate upon the physiognomic [1] of the kinds of animal *living*, the kinds of animal *soul*, leaving the systematic of bodily

---

[1] *Decline of the West*, English edition, Vol. I, pp. 99–103.

structure to the zoologists. For thereupon a wholly different *hierarchy*, one of life and not of body, discloses itself.

A plant lives, although only in the restricted sense a living being.[1] Actually there is life *in it*, or about it. " It " breathes, " it " feeds, " it " multiplies, we say, but in reality it is merely the *theatre* of processes that form one unity along with the processes of the natural environment, such as day and night, sunshine and soil-fermentation, so that the plant itself cannot will or choose. Everything takes place with it and in it. It selects neither its position, nor its nourishment, nor the other plants with which it produces its offspring. It does not move itself, but is moved by wind and warmth and light.

Above this grade of life now rises the freely mobile life of the animals. But of this there are *two stages*. There is one kind, represented in every anatomical genus from unicellular animals to aquatic birds and ungulates, whose living depends for its maintenance upon the *immobile* plant-world, for plants cannot flee or defend themselves. But above this there is a second kind, which lives on other animals and *whose living consists in killing*. Here

---

[1] *Decline of the West*, English edition, Vol. II, pp. 3 et seq.

the prey is itself mobile, and highly so, and moreover it is combative and well equipped with dodges of all sorts. This second kind also is found in all the genera of the system. Every drop of water is a battlefield and we, who have the land-battle so constantly before our eyes that it is taken for granted or even forgotten, shudder to see how the fantastic forms of the dead sea carry on the life of killing and being killed.

The animal of prey is the highest form of mobile life. It implies a maximum of freedom for self against others, of responsibility to self, of singleness of self, an extreme of necessity where that self can hold its own only by *fighting and winning and destroying.* It imparts a high dignity to Man, as a type, that he is a beast of prey.

A herbivore is by its destiny a *prey*, and it seeks to escape this destiny by flight, but beasts of prey must *get* prey. The one type of life is of its innermost essence defensive, the other offensive, hard, cruel, destructive. The difference appears even in the tactics of movement — on the one hand the habit of flight, fleetness, cutting of corners, evasion, concealment, and on the other the *straight-line* motion of the attack, the lion's spring, the eagle's swoop. There are dodges and counter-dodges alike in the style of the strong and in that of the weak. Cleverness in the human sense, *active* cleverness, belongs only to beasts of prey. Herbivores are by comparison stupid, and not merely the "innocent" dove and the elephant, but even the noblest sorts like the bull, the horse, and the deer; only in blind rage or sexual excitement are these capable of fighting; otherwise they will allow themselves to be tamed, and a child can lead them.

Besides these differences in kind of motion, there are others, still more effective, in the organs of sense. For these are accompanied by differences in the mode of apprehending, of having, a "world." In itself every being lives in Nature, in an environment, irrespective of whether it notices this environment, or is noticeable in it, or neither. But it is the relation — mysterious, inexplicable by any human reasoning — that is established between animal and environment by touching, ordering, and understanding, which creates out of mere environment a *world-around*. The higher herbivores are ruled by the ear, but above all by *scent*;[1] the higher carnivores on the other hand *rule with the eye*. Scent is the characteristically defensive sense. The nose catches the quarter and the distance of danger

---

[1] v. Üxküll *Biologische Weltanschauung* (1913), pp. 67 et. seq.

and so gives the flight-movement the appropriate direction, *away from* something.

But the eye of the preying animal gives a *target.* The very fact that, in the great carnivores as in man, the two eyes can be fixed on one point in the environment enables the animal to bind its prey. In that hostile glare there is already implicit for the victim the doom that it cannot escape, the spring that is instantly to follow. But this act of fixation by two eyes disposed forward and parallel is equivalent to the birth *of the world,* in the sense that Man possesses it — that is, as a picture, as a world before the eyes, as a world not merely of lights and colours, but of perspective distance, of space and motions in space, and of objects situated at definite points. This way of seeing which all the higher carnivores possess — in herbivores, e.g. ungulates, the eyes are set sideways, each giving a different and non-perspective impression — implies in itself the notion of *commanding.* The world-picture is the environment as *commanded* by the eyes. The eye of the beast of prey determines things according to position and distance. It apprehends the horizon. It measures up in this *battlefield* the objects and conditions of attack. Sniffing and spying, the way of the hind and the way of the falcon, are related as slavery and dominance. There is an infinite sense of power in this quiet wide-angle vision, a feeling of freedom that has its source in *superiority*, and its foundations in the knowledge of greater strength and consequent certainty of being no one's prey. The *world* is the prey, and in the last analysis it is owing to this fact that human culture has come into existence.

And, lastly, this fact of an innate superiority has become intensified, not only outwards, with respect to the light-world and its endless distances, but also inwards, as regards the sort of soul that the strong animals possess. The *soul* — this enigmatic something which we feel when we hear the word used, but of which the essence baffles all science, the divine spark in this living body which in this divinely cruel, divinely indifferent world has either to rule or to submit — is the *counter-pole* of the light-world about us, and hence man's thought and feeling are very ready to assume the existence of a world-soul in it. The more solitary the being and the more resolute it is in forming its own world against all other conjunctures of worlds in the environment, the more definite and strong the cast of its soul. What is the opposite of the soul of a lion? The soul of a cow. For strength of individual soul the herbivores substitute numbers, the herd, the common feeling and doing of masses. But the less one needs others, the more powerful one is. A beast of prey is everyone's foe. Never does he tolerate an equal in his den. Here we are at the root of the truly royal idea of *property*. Property is the domain in which one exercises unlimited power, the power that one has gained in battling, defended against one's peers, victoriously upheld. It is not a right to mere having, but the sovereign right to do as one will with one's own.

Once this is understood, we see that there are carnivore and there are herbivore *ethics*. It is beyond anyone's power to alter this. It pertains to the inward form, meaning, and tactics of all life. It is simply a *fact*. We can annihilate life, but we cannot alter it in kind. A beast of prey tamed and in captivity — every zoological garden can furnish examples — is mutilated, world-sick, inwardly dead. Some of them voluntarily hunger-strike when they are captured. Herbivores give up nothing in being domesticated.

Such is the difference between the destiny of herbivores and that of the beast of prey. The one destiny only menaces, the other enhances as well. The former depresses, makes mean and cowardly, while the latter elevates through power and victory, pride and hate. The former is a destiny that is imposed on one, the latter a destiny that is identical with oneself. And the fight of nature-within against nature-without is thus seen to be, not *misery*, as Schopenhauer and as Darwin's "struggle for existence" regard it, but a grand meaning that *ennobles* life, the *amor fati* of Nietzsche. And it is to this kind and not the other that Man belongs.

FOUR

MAN is no simpleton, "naturally good" and stupid, not a semi-ape with technical tendencies, as Haeckel describes him and Gabriel Max portrays him.[1] Over these pictures there still falls the plebeian shadow of Rousseau. No, the tactics of his living are those of a splendid beast of prey, brave, crafty, and cruel. He lives by attacking and killing and destroying. He wills, and has willed ever since he existed, to be master.

Does this mean, however, that technics is actually older than man? Certainly not. There is a vast difference between man and all other animals. The

---

[1] It is only the enraged demon of classification that haunts the pure anatomists which brought man close to the apes; moreover, even by them it is today coming to be regarded as an overhasty and shallow conclusion; see, for instance, Klaatsch, himself a Darwinian (*Der Werdegang der Menschheit*, 1920), pp. 29 et seq. For in the very "system" itself man stands off the line and outside all ordering — very primitive in many parts of his bodily structure and freakish in others. But that does not concern us here. It is his life we are studying, and in his destiny, *his soul*, he is an animal of prey.

technique of the latter is a *generic technique*. It is neither inventive nor capable of development. The bee type, ever since it existed, has built its honeycombs exactly as it does now, and will continue to build them so till it is extinct. They belong to it as the form of its wing and the colouring of its body belong to it. Distinctions between bodily structure and way of life are only anatomists' distinctions; if we start from the inner form of the life instead of that of the body, tactics of living and organization of body appear as one and the same, both being expressions of *one* organic actuality. "Genus" is a form, not of the visible and static, but of mobility — a form, not of so-being, but of so-doing. Bodily form is the form of the *active* body.

Bees, termites, beavers build wonderful structures. Ants know agriculture, road-making, slavery, and war-management. Nursing, fortification, organized migration are found widely spread. All that man can do, one or another sort of animal has achieved. Free-moving life in general contains tendencies that exist, dormant, as *potentialities*. Man achieves nothing that is not achievable by *life as a whole*.

And yet — all this has at bottom nothing whatever to do with human technics. This generic

technique is *unalterable;* that is what the word
" instinct " means. Animal " thought," being strictly
connected with the immediate here-and-now and
knowing neither past nor future, knows also neither
experience nor anxiety. It is not true that the female animal "cares" for her young. Care is a
feeling that presupposes mental vision into the
future, concern for what *is to be,* just as regret presupposes knowledge of what *was.* An animal can
neither hate nor despair. Its nursing activity is, like
everything else above mentioned, a dark unconscious response to impulse such as is found in many
types of life. It is a property of the species and not
of the individual. Generic technique is not merely
unalterable, but also *impersonal.*

The unique fact about human technics, on the
contrary, is that it is *independent* of the life of the
human genus. It is the one instance in all the history
of life in which the individual frees himself from
the compulsion of the genus. One has to meditate
long upon this thought if one is to grasp its immense
implications. Technics in man's life is conscious,
arbitrary, alterable, personal, *inventive.* It is
learned and improved. Man has become the *creator*
of his tactics of living — that is his grandeur and
his doom. And the inner form of this creativeness

we call culture — to be cultured, to cultivate, to suffer from culture. The man's creations are the expression of this being in *personal* form.

Chapter III

THE ORIGIN OF MAN: HAND AND TOOL

FIVE

SINCE when has this type of the *inventive carnivore* existed? Or, what comes to the same thing, since when have there been men? What is man? And how did he come to be man?

The answer is — through the genesis of the hand. Here is a weapon unparalleled in the world of free-moving life. Compare with it the paw, the beak, the horns, teeth, and tail-fins of other creatures. To begin with, the sense of touch is concentrated in it to such a degree that it can almost be called the organ of touch, in the sense that the eye is the organ of vision, and the ear of hearing. It distinguishes not only hot and cold, solid and liquid, hard and soft, but, above all, weight, form, and position of resistances, etc. — in short, *things in space*. But, over and above this function, the *activity* of living is gathered into it so completely that the whole bearing and allure of the body has — simultaneously — taken shape in accordance with it. There is nothing in the whole world that can be set beside this member, capable equally of touch and action. To the eye of the beast of prey which regards the world "theoretically" is added the hand of man which commands it *practically*.

Its origin must have been *sudden*; in terms of the tempo of cosmic currents it must have happened, like everything else that is decisive in world-history (epoch-making, in the highest sense), as abruptly as a flash of lightning or an earthquake. Here again we have to emancipate ourselves from the nineteenth-century idea, based on Lyell's geological researches, of an "evolutionary" process. Such a slow, phlegmatic alteration is truly appropriate to the English nature, but it does not represent Nature. To support the theory, since measurable periods disclosed no such process, time has been flung about in millions of years. But in truth we cannot distinguish geological strata unless *catastrophes* of unknown kinds and sources have separated them for us, nor yet species of fossil creatures unless they appear suddenly and hold on unaltered till their extinction. Of the "ancestors" of Man we know nothing, in spite of all our research and comparative anatomy. The human skeleton has been, ever since it appeared, just what it is now — one can observe even the Neandertal type in any public gathering. It is impossible, therefore, that hand, upright gait, the position of the head, and so forth should have developed successively and independently. The whole thing hangs together and suddenly *is*.[1] World-history strides on from catastrophe to catastrophe, whether we can comprehend and prove the fact or not. Nowadays, since de Vries,[2] we call it Mutation. It is an inner change that suddenly seized all specimens of genus, of course " without rhyme or reason," like everything else in actuality. It is the mysterious rhythm of the real.

Further, not only must man's hand, gait, and posture have come into existence together, but — and this is a point that no one hitherto has observed — *hand and tool* also. The unarmed hand is in

---

[1] As to this "evolution" in general — the Darwinians say that the possession of so admirable a weapon favoured and preserved the species in the struggle for existence. But for the weapon to confer an advantage it must first be ready, and the unfinished weapon would be a useless burden, and so a positive disadvantage, during the course of its evolution — an evolution which, be it noted, has to be regarded as taking thousands of years. And how, is it imagined, did the process *start?* It is somewhat imbecile to hunt down causes and effects, which after all are forms of man's thinking and not of the world's becoming, in the hope of penetrating the secrets of that world.

[2] H. de Vries: *Die Mutationstheorie* (1901, 1903; English translation, 1910).

itself useless. It requires a weapon to become itself a weapon. As the implements took form from the shape of the hand, so also the *hand from the shape of the tool.* It is meaningless to attempt to divide the two chronologically. It is impossible that the formed hand was active, even for a short time, without the implement. The earliest remains of man and of his tools are equally old.

What has divided, however — not chronologically, but logically — is the technical *process*, so that the making and the using of the tool are different things. As there is a technique of violin-making and another of violin-playing, so there is a technique of shipbuilding and another of sailing, the bowyer's craft and the archer's skill. No other preying animal even *selects* its weapon, but man not only selects it, but makes it, and according to his own individual ideas. And with this he obtains a terrific superiority in the struggle with his own kind, with other beasts, and with Nature.

This is what constitutes his liberation from the compulsion of the genus, a phenomenon unique in the history of all life on this planet. With this, man *comes into being.* He has made his active life to a large extent free of the conditions of his body. The genus-instinct still perseveres in full strength, but there has detached itself the thought and intelligent action of the individual, which is independent of the genus. This freedom consists in freedom of choice. Everyone makes his own weapon, according to his own skill and his own reasoning. The vast hoards of mis-shaped and rejected pieces that we find are eloquent of the carefulness of this original thinking-doing.

If, nevertheless, these pieces are so similar that one can — though with doubtful justification — deduce the existence of distinguishable "cultures" such as Acheulean and Solutrean, and even postulate therefrom — this certainly without justification — time-parallels in all the five continents, the explanation lies in the fact that this liberation from the compulsion of the genus only emanated at first as a grand *possibility* and fell far short of any actualized individualism. No one likes to pose as a freak, nor on the other hand merely to imitate another. In fact, everyone thinks and works for himself, but the life of the genus is so powerful that in spite of this the product is everywhere similar — as it is, at bottom, even today.

So besides the "thought of the eye," the comprehending and keen glance of the great beasts of prey, we have now the "thought of the hand."

From the former in the mean time has developed the thought that is theoretical, observant, contemplative — our " reflection " and " wisdom " — and now from the latter comes the practical, active thought, our " cunning " and " intelligence " proper. The eye seeks out cause and effect, the hand works on the principle of means and end. The question of whether something is suitable or unsuitable — the criterion of the *doer* — has nothing to do with that of true and false, the values of the *observer*. And an aim is a *fact*, while a connexion of cause and effect is a *truth*.[1] In this wise arose the very different modes of thought of the truth-men — the priest, the scholar, the philosopher — and the fact-men — the statesman, the general, the merchant. Ever since then, today even, the commanding, directing, clenching hand is the expression of a will, so much so that we have actually a graphology and a palmistry, not to mention figures of speech such as the " heavy hand " of the conqueror, the " dexterity " of the financier, and the " hand " revealed in the work of a criminal or an artist.

With his hand, his weapon, and his personal thinking man became *creative*. All that animals do

---

[1] *Decline of the West*, English edition, Vol. I, pp. 141 et seq.; Vol. II, pp. 212 et seq.

remains inside the limits of their genus-activity and does not enrich life at all. Man, however, the creative animal, has spread such a wealth of inventive thought and action all over the world that he seems perfectly entitled to call *his* brief history " world-history " and to regard his entourage as "humanity," with all the rest of Nature as a background, an object, and a means.

The act of the thinking hand we call a *deed*. There is already activity in the existence of the animals, but deeds begin only with Man. Nothing is more enlightening in this connexion than the story of fire. Man *sees* (cause and effect) how a fire starts, and so also do many of the beasts. But Man alone (end and means) *thinks out* a process of starting it. No other act so impresses us with the sense of creation as this one. One of the most uncanny, violent, enigmatic phenomena of Nature — lightning, forest fire, volcano — is henceforth called into life by Man himself, *against* Nature. What it must have been to man's soul, that first sight of a fire evoked by himself!

SIX

UNDER the mighty impress of this free and conscious *single act*, which thus emerges from the uniformity of the impulsive and collective genus-activity, the genuine human soul now forms — a very solitary soul (even as compared with those of the other beasts of prey) with the proud and pensive look of one knowing his own destiny, with unrestrained sense of power in the fist habituated to deeds, a foe to everyone, killing, *hating*, resolute to conquer or die. This soul is profounder and more passionate than that of any animal whatsoever. It stands in irreconcilable opposition to the whole world, from which its own creativeness has sundered it. It is the soul of an *upstart*.

Earliest man settled alone like a bird of prey. If several " families " drew together into a pack, it was a pack of the loosest sort. As yet there is no thought of tribes, let alone peoples. The pack is a chance assembly of a few males, who for once do not fight one another, with their women and the children of their women, without communal feeling and wholly free. They are not a "we" like the mere herd of specimens of a genus.

The soul of these strong solitaries is warlike through and through, mistrustful, jealous of its own power and booty. It knows the intoxication of feeling when the knife pierces the hostile body, and the smell of blood and the sense of amazement strike together upon the exultant soul. Every real "man," even in the cities of Late periods in the Cultures, feels in himself from time to time the sleeping fires of this primitive soul. Nothing here of the pitiful estimation of things as "useful" or "labour-saving," and less still of the toothless feeling of sympathy and reconciliation and yearning for quiet. But instead of these the full pride of knowing oneself feared, admired, and hated for one's fortune and strength, and the urge to vengeance upon all, whether living beings or things, that constitute, if only by their mere existence, a threat to this pride.

And this soul strides forward in an ever-increasing alienation from *all* Nature. The weapons of the beasts of prey are natural, but the armed fist of man with its artificially made, thought-out, and selected weapon is not. *Here begins "Art" as a counter-concept to " Nature."* Every technical process of man is an art and is always so described — so, for instance, archery and equitation, the art of war, the arts of building and government, of sacrificing and prophesying, of painting and versification, of scientific experiment. Every work of man is artificial, unnatural, from the lighting of a fire to the achievements that are specifically designated as " artistic " in the high Cultures. The privilege of creation has been wrested from Nature. " Free will " itself is an act of rebellion and nothing less. Creative man has stepped outside the bounds of Nature, and with every fresh creation he departs further and further from her, becomes more and more her enemy. *That* is his " world-history," the history of a steadily increasing, fateful rift between man's world and the universe — the history of a rebel that grows up to raise his hand against his mother.

This is the beginning of man's *tragedy* — for Nature is the stronger of the two. Man remains dependent on her, for in spite of everything she embraces him, like all else, within herself. All the great Cultures are *defeats*. Whole races remain, inwardly destroyed and broken, fallen into barrenness and spiritual decay, as corpses on the field.

The fight against Nature is hopeless and yet — it will be fought out to the bitter end.

Chapter IV

THE SECOND STAGE: SPEECH AND ENTERPRISE

SEVEN

How long the age of the armed hand lasted — that is, since when man has been man — we do not know. In any case the total of years does not matter, although today we still set it far too high. It is not a matter of millions of years, nor even of several hundreds of thousands. Nevertheless a considerable number of millennia must have flowed away.

But now comes a second epoch-making change, as abrupt and as immense as the first, and like it transforming man's destiny from the foundations — once more a true " mutation " in the sense above indicated. Prehistoric archæology observed this long ago, and in fact the things that lie in our museums do suddenly begin to look different. Clay vessels appear, and traces of " agriculture " and " cattle-breeding " (though this is a rash use of terms that connote something much more modern), hut-building, graves, indications of travel. A new world of technical ideas and processes sets in. The museum standpoint, which is far too superficial and obsessed with the mere ordering of finds, has differentiated older and newer Stone Ages, the Palæolithic and Neolithic. This nineteenth-century classification has long been regarded with uncomfortable doubts, and in the last few decades attempts have been made to replace it by something else. But scholars are still sticking to the idea of classifying *objects* (as terms like Mesolithic, Miolithic, Mixolithic indicate) and hence they are getting no further. What changed was, not equipment, but *Man*. Once more, it is only from his soul that man's history can be discovered.

The date of this mutation can be fixed with fair accuracy as being somewhere in the fifth millennium before Christ.[1] Two thousand years later at most, the high Cultures are beginning in Egypt and Mesopotamia. Truly the tempo of history is working up tragically. Hitherto thousands of years have scarcely mattered at all, but now every century becomes important. With tearing leaps, the rolling stone is approaching the abyss.

---

[1] Following the researches of De Geers on Swedish banded clay (*Reallexikon der Vorgeschichte*, Vol. II, article "*Diluvialchronologie.*"

But what in fact has happened? If one goes more deeply into this new form-world of man's activities, one soon perceives most bizarre and complicated linkages. These techniques, one and all, presuppose one another's existence. The keeping of tame animals demands the cultivation of forage stuffs, the sowing and reaping of food-plants require draught-animals and beasts of burden to be available, and these, again, the construction of pens. Every sort of building requires the preparation and transport of materials, and transport, again, roads and pack-animals and boats.

What in all this is the *spiritual* transformation? The answer I put forward is this — *collective doing by plan*. Hitherto each man had lived his own life, made his own weapons, followed his own tactics in the daily struggle. None needed another. This is what suddenly changes now. The new processes take up long periods of time, in some cases years — consider for instance the time that elapses between the felling of a tree and the sailing of the ship that is built out of it. The story divides itself into a set of well-arranged separate " acts " and a set of " plots " working out in parallel with one another. And for this collective procedure the indispensable prerequisite is a medium, *language*.

Speaking in sentences and words, therefore, cannot have begun either earlier or later, but must have come just then — quickly, like everything decisive, and, moreover, in close connexion with man's new methods. This can be proved.

What is "speaking"?[1] Indubitably a process having for its object the imparting of information, an activity that is practised continuously by a number of human beings amongst themselves. "Speech" or "language" is only an abstraction from this, the inner (grammatical) form of speaking, and therewith of words. This form must be common property and must have a certain permanence if information is really to be imparted by its means. I have elsewhere shown that speaking in sentences is preceded by simpler forms of communication, such as signs for the eye, signals, gestures, warning and threatening calls. All these continue in use, even today, as auxiliaries to speaking, as melodious speech, emphasis, play of features and hands, and (in written speech) punctuation.

Nevertheless, "fluent" speaking is, by reason of its content, something quite new. Ever since Hamann and Herder, men have been setting themselves the question of its origin. But if all answers so far

---

[1] See *Decline of the West*, English edition, Vol. II, ch. v.

have been more or less unsatisfactory, it is because the *intention* of the question has been wrong. For the origin of speaking in words is not to be found in the activity of speaking itself. That was the error of the Romantics, who (remote as ever from actuality) deduced speech from the "primary poesy" of mankind. Nay, more, they thought that speech was itself this poesy — myth, lyric, and prayer rolled into one — and that prose was merely something later and degraded for common everyday use. But had this been so, the inner form of the speaking, the grammar, the logical build of the sentence, would have borne quite a different look; in reality it is precisely the very primitive languages (such as those of Bantu and Turcoman tribes) that show most emphatically the tendency to mark differences clearly, sharply, and unmistakably.[1]

This, in turn, brings us to the fundamental error of those sworn foes of Romanticism, the rationalists, who are for ever chasing the idea that what the sentence expresses is a *judgment* or a *thought*. They sit at their writing-tables, surrounded by books, and

---

[1] So much so that in many tongues the sentence is a single monstrous word, in which everything that is intended to be said is expressed by means of syllables of classification prefixed and suffixed according to rule.

research into the minutiæ of *their own* thoughts and writings. Consequently the "thought" appears to them as the *object* of the speaking, and (since usually they sit alone) they forget that beyond the speaking there is a hearing, beyond a question an answer, beyond an Ego a Tu. They say "speech," but what they mean is the oration, the lecture, the discourse. Their view of the origin of speech is, therefore, false, for they look upon it as *monologue*.

The correct way of putting the question is not how, but *when* did speaking in words come into existence? And once this question takes this form, all very soon becomes clear. The object of speaking in sentences, usually misunderstood or ignored, is settled by the period in which it became customary to speak thus (that is, "fluently"), and displayed quite clearly in the form of sentence-building. Speaking did not arise by way of monologue, nor sentences by way of oratory; the source is in the *conversation* of several persons. The object is not one of understanding as a consequence of reflection, but one of reciprocal understanding as a consequence of question and answer. What, then, are the basic forms of the speaking? Not the judgment and declaration, but the *command*, the expression of obedience, the enunciation, the question, the affirmation or negation. These are sentences, originally quite brief, which are *invariably* addressed to others, such as "Do this!" "Ready?" "Yes!" "Go ahead!" Words as designations of notions [1] are only products of the *object* of the sentence, and hence it is that the vocabulary of a hunting tribe is from the outset different from that of a village of cowherds or a seafaring coastal population. Originally, speaking was a difficult activity,[2] and it may be assumed that it was limited to bare essentials. Even today the peasant is slow of speech as compared with the townsman — who is so accustomed to speaking that he cannot hold his tongue and must, from mere boredom, chatter and make conversation as soon as he has nothing else to do, whether he has anything really to say or not.

The original *object* of speech is the *carrying-out of an act* in accordance with intention, time, place, and means. Clear and unequivocal construction is therefore the first essential, and the difficulty of both conveying one's meaning to, and imposing one's will on, another produced the technique of grammar,

---

[1] A notion is an ordering of things, situations, activities, in classes of *practical* generality. The horse-breeder does not say "horse," but "grey mare" or "bay foal"; the hunter says, not "wild boar," but "tusker," "two-year-old," "shoat."

[2] Certainly it would only have been adults who could speak fluently, just as was the case far later on with writing.

sentences, and constructions, the correct modes of ordering, questioning, and answering, and the building-up of classes of words — on the basis of *practical* and not theoretical intentions and purposes. The part played by theoretical reflectiveness in the beginnings of speaking in sentences was practically *nil*. All speech was of a practical nature and proceeded from the "thought of the hand."

EIGHT

A "COLLECTIVE doing by plan" may be more briefly called an *enterprise*. *Speech* and *enterprise* stand in precisely the same relation to each other as the older pair *hand* and *implement*. Speaking to several persons developed its inner, grammatical form in the practice of carrying out jobs, and *vice versa* the habit of doing jobs got its schooling from the methods of a thinking that had to work with words. For speaking consists in imparting something to another's thought. If speaking is an act, it is an *intellectual* act with *sensuous* means. Very soon it no longer needed the original immediate connexion with physical doing. The epoch-making innovation of the fifth millennium B.C. was, in fact, that thereafter the thinking, the intellect, the reason, that which (call it by what name you please) had been emancipated by speech from dependence upon the doing *hand* proceeded to set itself up against Soul and Life *as a power in itself*.

The purely intellectual thinking-over, the "calculation," which emerges at this point — sudden, decisive, and radical — amounts to this, that collective doing is as effectively a *unit* as if it were the doing of some single giant. Or as Mephistopheles ironically says to Faust:

> ". . . when to my car
> my money yokes six spankers, are
> their limbs not my limbs? Is't not I
> on the proud racecourse that dash by?
> Mine all the forces I combine,
> the four-and-twenty legs are mine!"[1]

Man, the preying animal, insists *consciously* on increasing his superiority far beyond the limits of his bodily powers. To this will-to-more-power of his he even sacrifices an important element of *his own* life. The thought of, the calculation for, intenser effectiveness comes first, and for the sake of it he is quite willing to give up a little of his personal freedom. Inwardly, indeed, he remains independent. But history does not permit one step to be taken back. Time, and therefore Life, are irreversible. Once habituated to the collective doing

---

[1] Anster's translation.—*Tr.*

and its successes, man commits himself more and more deeply to its fateful implications. The enterprise in the mind requires a firmer and firmer hold on the life of the soul. Man has become the slave of his thought.

The step from the use of personal tools to the common enterprise involves an immensely increased *artificiality* of procedure. The mere working with artificial *material* (as in pottery, weaving, and matting) does not as yet mean a great deal, although even it is something much more intelligent and *creative* than anything before it. But traces have come down to us of some few processes, standing far above the many of more ordinary kinds of which today we can know nothing, which presuppose very great powers of thought indeed. Above all, those which grew out of the *idea* of *building*. Long before there was any knowledge of metals, there were flint mines in Belgium, England, Austria, Sicily, and Portugal — complete with shafts and galleries, ventilation and drainage, and tools fashioned of deerhorn — which certainly go back to these times.[1] In the early Neolithic period Portugal and northwestern Spain had close relations with Brittany (passing *round* southern France), and Brittany in

---

[1] *Reallexikon der Vorgeschichte*, Vol. I, article "*Bergbau.*"

turn with Ireland, which presupposes regular navigation and, therefore, the building of seaworthy ships of some sort, though we know nothing about these. There are megaliths in Spain built of hewn stones of vast size, with cap-stones weighing more than a hundred tons, which must have been brought from great distances and placed in position somehow, though again we know nothing of the technique employed. In truth, have we any clear notion of how much thought, consultation, superintendence, ordering was required, over months and years on end, for the quarrying and transport of this material, for the assignment of tasks in time and in space, the planning, the undertaking and execution of such work? How compare the prolonged forethought necessary for such transportation across the open sea with the production of a flint knife? Even the " composite bow " which appears in Spanish rock-pictures of the period demands for its construction sinews, horn, and special woods, all from different sources, as well as a complicated process of manufacture that took five to seven years. And the " discovery," as we so naïvely call it, of the wagon — how much thinking, ordering, and doing it presupposes, ranging from the determination of the purpose and kind of movement required, the choice and preparation of the *road* (a point usually ignored), and the provision or breaking-in of draught-animals, to consideration of the bulk, weight, and lashing of the load, the management and housing of the convoy!

Another and quite different world of creations arises out of the "thought" of procreation — namely, the *breeding* of plants and animals, in which man himself takes the place of Nature the creatrix, imitates her, modifies her, improves on her, overrides her. From the time when he began *cultivating* instead of gathering plants, there is no doubt that he consciously modified them for his own ends. At any rate the specimens discovered belong to species that have never been found in a wild state. And even in the oldest finds of animal bones that indicate cattle-keeping of any sort, we perceive already the consequences of "domestication," which, partly if not wholly, must have been intentional and brought about by deliberate breeding.[1] The prey-idea of the carnivore at once widens and includes not only the slain victims of the hunt, but also the free cattle[2] that graze freely within (or even

---

[1] Hilzheimer: *Natürliche Rassengeschichte der Haussäugetiere* (1926).

[2] In the same conditions as the live-stock of our woods today.

without) a man-made hedge.[1] They belong to someone — a clan, a hunting-group — and the owner will fight to maintain his right of exploitation. The capture of animals for breeding-purposes, which presupposes the cultivation of foodstuffs for them, is only one of many modes of possession then practised.

I have already shown that the birth of the armed hand had had as its result a *logical* separation of two techniques — namely, that of making and that of using the weapon. Similarly, the verbally managed enterprise led to the separation of the activities of thought from those of the hand. In every enterprise *planning out* and *carrying out* are distinct elements, and, as between these, practical thought henceforth takes the leading part. There is *director's work* and there is *executant's work,* and this fact has been the basic technical form of all human life ever since. Whether it is a matter of hunting big game or building temples, an enterprise of war or of rural development, the founding of a firm or of a state, a caravan journey or a re-

---

[1] Even in the nineteenth century the Indian tribes still followed the great buffalo-herds, just as the Gauchos of Argentina follow the privately owned cattle-herds today. Thus, in certain cases, we find nomadism growing *out* of settlement and not the other way about.

bellion or even a crime — always the first prerequisite is an enterprising, inventive head to conceive the idea and direct the execution, to command and to allot the rôles — in a word, someone who is born to be a leader of others who are not so.

For in this age of verbally managed enterprises there are not only two sorts of technics — these, by the way, diverging more and more definitely as the centuries go on — but also *two kinds of men*, differentiated by the fact of their talent lying in one or in the other direction. As in every process there is a technique of direction and a technique of execution, so, equally self-evidently, there are *men whose nature is to command and men whose nature is to obey, subjects and objects of the political or economic process in question*. This is the basic form of the human life that since the change has assumed so many and various shapes, and it is only to be eliminated along with life itself.

Admittedly this is artificial, contrary to Nature — but that is just what " Culture " *is*. Fate may ordain, and at times does ordain, that man should imagine himself able to abolish it — *artificially* — but nevertheless it is unshakably a *fact*. Governing, deciding, guiding, commanding is an art, a difficult technique, and like any other it presupposes an innate talent. Only children imagine that a king goes to bed with his crown, and only sub-men of the monster-city, Marxists and literary people, imagine the same sort of thing about business kings. Undertaking is *work*, and it is only as the result of that work that the manual labour became possible. Similarly the discovery, thinking-out, calculation, and management of new processes is a *creative* activity of gifted heads, and the executive rôle falls to the uncreative as a necessary consequence. And here we meet an old friend, now a little out of date, the question of genius and talent. Genius is — literally [1] — creative power, the divine spark in the individual life that in the stream of the generations mysteriously and suddenly appears, is extinguished, and a generation later reappears with equal suddenness. Talent is a gift for particular tasks already there, which can be developed by tradition, teaching, training, and practice to high effectiveness. Talent in its exercise presupposes genius — and not *vice versa*.

Finally, there is a natural distinction of grade between men born to command and men born to service, between the leaders and the led of *life*. The

---

[1] It comes from the Latin *genius*, the masculine generative force.

existence of this distinction is a plain *fact*, and in healthy periods and by healthy peoples it is admitted (even if unwillingly) by everyone. In the centuries of decadence the majority force themselves to deny or to ignore it, but the very insistence on the formula that " all men are equal " shows that there is something here that has to be explained away.

NINE

NOW, this verbally managed enterprise involves an immense loss of freedom — the old freedom of the beast of prey — *for the leader and the led alike.* They *both* become intellectual, spiritual members of a higher unit, body and soul. This we call *organization*, the gathering of active life into definite forms, into the condition of being " in form " for the enterprise, whatever it may be. With collective doing the decisive step is taken from *organic* to *organized* existence, from living in natural to living in artificial groupings, from the pack to the people, the tribe, the social class, the State.

And out of the combats of individual carnivores there has sprung war, as an *enterprise* of tribe against tribe, with leaders and followers, with organized marches, surprises, and actions. Out of the annihilation of the vanquished springs the *law* that is imposed upon the vanquished. Human law is ever a law of the stronger to which the weaker must conform,[1] and this law, considered as something permanently valid between tribes, constitutes "*peace.*" Such a peace also prevails within the tribe, so that its forces may be available for action outwards; *the State is the internal order of a people for its external purpose.* The State is as form, as possibility, what the history of a people is as actuality. But history, of old as now, is war-history. Politics is only a temporary substitute for war that uses more intellectual weapons. And the male part of a community is originally synonymous with its *host.* The character of the free beast of prey passes over, in its essential features, from the individual to the organized people, the animal with one soul and many hands.[2] The technics of government, war, and diplomacy have all this same root and have in all ages a profound inward relationship with each other.

There are peoples whose strong breed has kept the character of the beast of prey, seizing, conquering, and lording peoples, lovers of the fight against *men,* who leave the economic fight against Nature to others, whom in due course they plunder and

---

[1] *Decline of the West,* English edition, Vol. II, pp. 64 et seq.
[2] And, be it added, *one* head, and not many.

subject. Piracy is as old as navigation, the raiding of the trade-route as old as nomadism, and wherever there is peasantry there is enslavement to a warlike nobility.

For with the organization of undertakings comes the separation of the political and the economic sides of life, that directed towards *power* and that directed towards *booty*. We find not merely an *internal* articulation of the people according to activities — warriors and workers, chiefs and peasants — but also the organization of whole tribes for a single economic occupation. Even then there must have been hunting, cattle-breeding, and agricultural tribes, mining, pottery, and fishing villages, political organizations of seafarers and traders — and over and above these, conqueror peoples *without* economic occupation at all. The harder the battle for power and booty, the closer and stricter the bonding of the individuals by law and force.

In the tribes of this primitive sort the individual life mattered little or nothing. For consider that in every sea voyage (the Icelandic sagas are illuminating here) only a proportion of the ships reached port, that in every great building task no small part of the workmen perished, that whole tribes starved in time of drought — clearly, all that mattered was that enough were left to represent the *spirit* of the whole. The numbers decreased rapidly, but what was felt as annihilation was, not the loss of one or even of many, but *the extinction of the organization, of the " we."*

In this increasing interdependence lies the quiet and deep revenge of Nature upon the being that has wrested from her the privilege of creation. This petty creator *against* Nature, this revolutionary in the world of life, has become the slave of his creature. The Culture, the aggregate of artificial, personal, self-made life-forms, develops into a close-barred cage for these souls that would not be restrained. The beast of prey, who made others his domestic animals in order to exploit them, has taken himself captive. The great symbol of this fact is the human *house.*

That, and his increasing numbers, in which the individual disappears as unimportant. For it is one of the most fateful consequences of the human spirit of enterprise that the population multiplies. Where anciently the pack of a few hundreds roamed, there is sitting a people of tens of thousands.[1] There are scarcely any regions empty of men. People borders on people, and the mere *fact* of the frontier — the

---

[1] Nay, today, there is *squeezed* one of millions.

limit of one's own power — arouses the old instincts to hate, to attack, to annihilate. The frontier, of whatever kind it may be, even the intellectual frontier, is the mortal foe of the Will-to-Power.

It is not true that human technics saves labour. For it is an essential characteristic of the personal and modifiable technics of man, in contrast to genus-technics, that every discovery contains the possibility and *necessity* of new discoveries, every fulfilled wish awakens a thousand more, every triumph over Nature incites to yet others. The soul of this beast of prey is ever hungry, his will never satisfied — that is the curse that lies upon this kind of life, but also the grandeur inherent in its destiny. It is precisely its best specimens that know least of quiet, happiness, or enjoyment. And no discoverer has ever accurately foreseen the *practical* effect of his act. The more fruitful the leader's work, the greater the need of executive hands. And so, instead of killing the prisoners taken from hostile tribes, men begin to enslave them, so as to exploit their bodily strength. This is the origin of Slavery, which must, therefore, be precisely as old as the slavery of domestic animals.

In general, these peoples and tribes multiply, so to say, *downwards*. What grows is not the number of "heads," but that of hands. The group of leader-natures *remains* small. It is, in fact, the pack of the true beasts of prey, the pack of the gifted who dispose, in one way or another, of the increasing *herd* of the others.

But even this lordship of the few is far removed from the ancient freedom — witness Frederick the Great's saying: " I am the first servant of my State." Hence the desperate efforts of the "exceptional" man to keep himself inwardly free. Here, and only here, begins *the individualism that is a reaction against the psychology of the mass.* It is the last uprising of the carnivore soul against its captivity behind the bars of the Culture, the last attempt to shake off the spiritual and intellectual *limitations* that are produced by, and represented by, the fact of large numbers. Hence the life-types of the conqueror, the adventurer, the hermit, and even certain types of criminals and Bohemians. The wished-for escape from absorption by the large number takes various forms — lordship over it, flight from it, contempt for it. The idea of personality, in its dark beginnings, is a protest against humanity in the mass, and the tension between these grows and grows to its tragic finale.

Hate, the most genuine of all race-feelings in the

beast of prey, presupposes *respect* for the adversary. A certain recognition of like spiritual rank is inherent in it. Beings that stand lower one *despises*. Beings that themselves stand low are *envious*. All primitive folk-tales, god-myths, and hero-sagas are full of such motives. The eagle hates only his peers, envies none, despises many and indeed all. Contempt looks downwards from the heights, envy peers upwards from below — and these two are the *world-historical* feelings of mankind organized in state and classes, whose (forcedly) peaceful specimens helplessly rattle the bars of the cage in which they are confined *together*. From this fact and its consequences *nothing* can liberate them. So it was and so it will be — or nothing at all will be. It has a significance, this fact of respect and contempt. To *alter* it is impossible. The destiny of man is pursuing its course and must accomplish itself.

Chapter V

THE LAST ACT: RISE AND END OF THE MACHINE CULTURE

TEN

THE culture of the armed hand had a long wind and got a grip on the whole genus man. The Cultures of speech and enterprise — we are at once in the plural, and several can be distinguished — in which personality and mass begin to be in spiritual opposition, in which the spirit becomes avid of power and lays violent hands on life, these Cultures embraced even at their full only a *part* of mankind, and they are today, after a few millennia, all extinguished and replaced. What we call "nature-peoples" and "primitives" are merely the remains of their living material, the ruins of forms that once were permeated with soul, cinders out of which the glow of becoming and departing has gone.

On this soil, from 3000 B.C. onwards, there now grew up, here and there, the *high Cultures*,[1] Cultures in the narrowest and grandest sense, each filling

---

[1] *Decline of the West*, English edition, Vol. I, pp. 103 et seq.

but a very small portion of the earth's space and each enduring for hardly a thousand years. The tempo is that of the final catastrophes. Every decade has significance, every year, almost, its special "look." It is world-history in the most genuine and most exacting sense. This group of passionate life-courses invented for its symbol and its "world" the *city*, in contrast to the village of the previous stage — the stone city in which is housed a quite artificial living, that has become divorced from mother earth and is *completely* anti-natural — the city of rootless thought, that draws the streams of life from the land and uses them up into itself.[1]

There arises "society"[2] with its hierarchy of classes, noble, priest, and burgher, as an *artificial* gradation of life against the background of "mere" peasantry — for the *natural* divisions are those of strong and weak, clever and stupid — and as the seat of a cultural evolution that is wholly intellectualized. There "luxury" and "wealth" reign. These are concepts which those who do not share them enviously misunderstand. For what is luxury but Culture in its most exacting form? Consider the

---

[1] *Decline of the West*, English edition, Vol. II, ch. iv, "The Soul of the City."

[2] *Decline of the West*, English edition, Vol. II, pp. 327 et seq., 343 et seq.

Athens of Pericles, the Baghdad of Haroun-al-Raschid, the Rococo. This urban Culture is luxury through and through, in all grades and callings, artificial from top to bottom, an affair of arts, whether arts of diplomacy or living, of adornment or writing or thought. Without an economic wealth that is concentrated in a few hands, there can be no " wealth " of art, of thought, of elegance, not to speak of the luxury of possessing a world-outlook, of thinking theoretically instead of practically. Economic impoverishment at once brings spiritual and artistic impoverishment in its train.

And, in this sense, the technical processes that mature in these Cultures are also spiritual luxuries, late, sweet, and fragile fruits of an increasing artificiality and intellectuality. They begin with the building of the tomb pyramids of Egypt and the Sumerian temple-towers of Babylonia, which come into being in the third millennium B.C., deep in the South, but signify no more than the victory over big *masses*. Then come the enterprises of Chinese, Indian, Classical, Arabian, and Mexican Cultures. And now, in the second millennium of our era, in the full North, there is our own Faustian Culture, which represents the victory of pure technical thought over big *problems*.

For these Cultures grow up, though *independently* of one another, yet in a series of which the sense is from South to North. The Faustian, west-European Culture is *probably* not the last, but *certainly* it is the most powerful, the most passionate, and — owing to the inward conflict between its comprehensive intellectuality and its profound spiritual disharmony — the most tragic of them all. It is possible that some belated straggler may follow it — for instance, a Culture may arise somewhere in the plains between the Vistula and the Amur — during the next millennium. But it is here, in our own, that the struggle between Nature and the Man whose historic destiny has made him pit himself against her is to all intents and purposes ended.

The Northern countryside, by the severity of the conditions of life in it — the cold, the continuous privation — has forged hard races, with intellects sharpened to the keenest, and the cold fires of an unrestrained passion for fighting, risking, thrusting forward — that which elsewhere [1] I have called *the passion of the Third Dimension.* There are, once more, beasts of prey whose inner forces struggle fruitlessly to break the superiority of thought, of

---

[1] *Decline of the West,* English edition, Vol. I, pp. 165 et seq., pp. 308 et seq.

organized artificial living, over the blood, to turn these into their servants, to elevate the destiny of the free personality to being the *very meaning* of the world. A will-to-power which laughs at all bounds of time and space, which indeed regards the boundless and endless as its specific target, subjects whole continents to itself, eventually embraces the world in the network of its forms of communication and intercourse, and *transforms* it by the force of its practical energy and the gigantic power of its technical processes.

At the beginning of every high Culture the two primary orders, nobility and priesthood — the beginnings of "society" — take shape clear of the peasant-life of the open land.[1] They are the embodiment of ideas, and, moreover, mutually exclusive ideas. The noble, warrior, adventurer lives in the world of *facts*, the priest, scholar, philosopher in his world of truths. The one is (or suffers) a *destiny*, the other thinks in *causality*. The one would make intellect the servant of a strong living, the other would subject his living to the service of the intellect. And nowhere has this opposition taken more irreconcilable forms than in the Faustian Culture, in which the proud blood of the beast

---

[1] *Decline of the West*, English edition, Vol. II, pp. 334 et seq.

of prey revolts for the last time against the tyranny of pure thought. From the conflict between the ideas of Empire and Papacy in the twelfth and thirteenth centuries to the conflict between the forces of a thoroughbred tradition — kingship, nobility, army — and the theories of a plebeian rationalism, liberalism, and socialism — from the French to the German revolution — history is one sequence of efforts to get the decision.

ELEVEN

THIS difference appears, in all its magnitude, in the contrast between the *Vikings of the blood and the Vikings of the mind* during the rise of the Faustian Culture. The first, thrusting insatiably out from the high North into the infinite, reached Spain in 796, Inner Russia in 859, Iceland in 861. In 861, too, Morocco was reached, and thence they ranged to Provence and the environs of Rome itself. In 865, by Kiev, the drive passed on to the Black Sea and Constantinople, in 880 to the Caspian, in 909 to Persia. They settled in Normandy and Iceland about 900, in Greenland about 980, and discovered North America about 1000. In 1029, from Normandy, they are in lower Italy and Sicily; in 1034, from Constantinople, they are in Greece and Asia Minor; and in 1066, from Normandy again, they conquer England.[1]

With a like boldness and a like hunger for

---

[1] K. Th. Strasser: *Wikinger und Normannen* (1928).

power and booty, in this case intellectual, Northern monks in the thirteenth and fourteenth centuries forced their way deep into the world of technical-physical problems. Here there was nothing of the idle and unpractical curiosity of the Chinese, Indian, Classical, and Arabian savants, no mere teleological speculation, no contemplative search for a picture of that which man cannot know. True, *every* scientific theory is a *myth* of the understanding about Nature's forces, and everyone is dependent, through and through, upon the religion with which it belongs.[1] But in the Faustian, and the Faustian alone, every theory is also from the outset a *working hypothesis*.[2] A working hypothesis need not be " correct," it is only required to be practical. It aims, not at embracing and unveiling the secrets of the world, but at making them serviceable to definite ends. Hence the advance in *mathematical* methods, due to the Englishmen Grosseteste (born 1175) and Roger Bacon (born *ca.* 1210), and the Germans Albertus Magnus (born 1193) and Witelo (born 1220). Hence, too, *experiment*, Bacon's " *Scientia experimentalis*," which is the interrogation of Nature under torture with the rack, lever,

---

[1] *Decline of the West*, English edition, Vol. II, ch. xi.
[2] *Decline of the West*, English edition, Vol. II, pp. 300 et seq.

and screw;[1] "*experimentum enim solum certificat,*" as Albertus Magnus put it. It is the stratagem of intellectual beasts of prey. They imagined that their desire was to "know God," and yet it was the forces of *inorganic* Nature — the invisible energy manifested in all that happens — that they strove to isolate, to seize, and to turn to account. This Faustian science, and it alone, is *Dynamics*, in contrast to the Statics of the Greeks and the Alchemy of the Arabs.[2] It is concerned, not with stuffs, but with forces. Mass itself is a function of energy. Grosseteste developed a theory of space as a function of light, Petrus Peregrinus a theory of magnetism. The Copernican theory of the earth's motion round the sun was foreshadowed in a manuscript of 1322 and formulated — more clearly and more profoundly than by Copernicus himself — by Oresme, who also anticipated the Galileian law of falling bodies and the Cartesian co-ordinate geometry. God was looked upon no longer as the Lord who rules the world from His throne, but as an infinite force (already imagined as almost impersonal) that is omnipresent in the world. It was a singular form of divine worship, this experimental

---

[1] *Decline of the West*, English edition, Vol. II, pp. 499 et seq.
[2] *Decline of the West*, English edition, Vol. I, pp. 380 et seq.

investigation of secret forces by pious monks. As an old German mystic said, " In thy serving of God, God serves thee."

Man, evidently, was tired of merely having plants and animals and slaves to serve him, and robbing Nature's treasures of metal and stone, wood and yarn, of managing her water in canals and wells, of breaking her resistances with ships and roads, bridges and tunnels and dams. Now he meant, not merely to plunder her of her materials, *but to enslave and harness her very forces* so as to multiply his own strength. This monstrous and unparalleled idea is as old as the Faustian Culture itself. Already in the tenth century we meet with technical constructions of a wholly new sort. Already the steam engine, the steamship, and the air machine are in the thoughts of Roger Bacon and Albertus Magnus. And many a monk busied himself in his cell with the idea of *Perpetual Motion*.[1]

This last idea never thereafter let go its hold on us, for success would mean the final victory over " God or Nature " (*Deus sive Natura*), a small world of one's own creation moving like the great world, in virtue of its own forces and obeying the

---

[1] *Decline of the West*, English edition, Vol. II, pp. 499 et seq. *Epistola de Magnete* of Petrus Peregrinus, 1269.

hand of man alone. To build a world *oneself,* to be *oneself* God — that is the Faustian inventor's dream, and from it has sprung all our designing and re-designing of machines to approximate as nearly as possible to the unattainable limit of perpetual motion. The booty-idea of the beast of prey is thought out to its logical end. Not this or that bit of the world, as when Prometheus stole fire, but the world itself, complete with its secret of force, is dragged away as spoil to be built into our Culture. But he who was not himself possessed by this will to power over all nature would necessarily feel all this as *devilish,* and in fact men have always regarded machines as the invention of the devil — with Roger Bacon begins the long line of scientists who suffer as magicians and heretics.

But the history of West European technics marched on. By 1500 the new Vikingism begun by Vasco da Gama and Columbus is under way. New realms are created or conquered in the East and West Indies, and a stream of Nordic blood [1] is poured out into America, where of old the

---

[1] For even the Spaniards, Portuguese, and French who went out thither must surely have been, for the most part, descendants of the barbarian conquerors of the Great Migrations. The remainder, that stayed behind, were of a human type that had already lasted out the Celts, the Romans, and the Saracens.

Icelandic seamen had set foot in vain. At the same time the Viking voyages of the intellect continued on a grand scale. Gunpowder and printing were discovered. From Copernicus and from Galileo on, technical processes followed one another thick and fast, all with the same object of extracting the inorganic forces from the world-around and making them, instead of men and animals, do the work.

With the growth of the towns, technics became *bourgeois.* The successor of those Gothic monks was the cultured lay inventor, the expert *priest of the machine.* Finally, with the coming of rationalism, the belief in technics almost becomes a materialistic religion. Technics is eternal and immortal like God the Father, it delivers mankind like God the Son, and it illumines us like God the Holy Ghost. And its worshipper is the progress-philistine of the modern age which runs from Lamettrie to Lenin.

In reality the passion of the inventor has *nothing whatever* to do with its consequences. It is his *personal* life-motive, his *personal* joy and sorrow. He wants to enjoy his triumph over difficult problems, and the wealth and fame that it brings him, for their own sake. Whether his discovery is useful or menacing, creative or distributive, he cares not a jot. Nor indeed is anyone in a position to know this in advance. The effect of a " technical achievement of mankind " is never foreseen — and, incidentally, " mankind " has never discovered anything whatever. Chemical discoveries like that of synthetic indigo and (what we shall presently witness) that of artificial rubber upset the living-conditions of whole countries. The electrical transmission of power and the discovery of the possibilities of energy from water have depreciated the old coal-areas of Europe *and their populations.* Have such considerations ever caused an inventor to suppress his discovery? Anyone who imagines this knows little of the beast-of-prey nature of man. All great discoveries and inventions spring from the delight of strong men in *victory*. They are expressions of personality and not of the utilitarian thinking of the masses, who are merely spectators of the event, but must take its consequences whatever they may be.

And these consequences are immense. The small band of born leaders, the undertakers and the inventors, force Nature to perform work that is measured in millions and thousands of millions of . . . horse-power, and in face of this the quantum of man's physical powers is so small as not to count. We understand the secrets of Nature as little as ever, but we do know the working hypothesis — not "true," but merely appropriate — which enables us to force her to *obey* the command that man expresses by the lightest touch on a switch or a lever. The pace of discovery grows fantastic, and withal — it must be repeated — human labour is *not* saved thereby. The number of necessary hands grows with the number of machines, since technical luxury surpasses every other sort of luxury,[1] and our artificial life becomes more and more artificial.

Since the discovery of the machine — the subtlest of all possible weapons against Nature — entrepreneurs and inventors have in principle devoted the number of hands that they needed to its *production*, the *working* of the machine being done by inorganic force — steam or gas pressure, electricity, heat — liberated from coal, petroleum, and water. But this difference has dangerously accentuated the spiritual tension between leaders and led. The two no longer understand each other. The earliest "enterprises" in the pre-Christian millennia required the *intelligent* co-operation of all concerned, who had to know and feel what it was all

---

[1] Compare the conditions of living of the working-classes in 1700 and in 1900, and in general the way of life of town workers as compared with those on the land.

about. There was, therefore, a sort of camaraderie in it, rather like that which we have today in sport. But even by the time of the great constructions of Babylonia and Egypt this cannot have been the case any longer. The individual labourers comprehended neither the object nor the purpose of the enterprise as a whole, to which they were indifferent and perhaps hostile. "Work" was a *curse*, as in the Biblical story of the Garden of Eden. And now, since the eighteenth century, innumerable "hands" work at things of which the real rôle in life (even as affecting themselves) is entirely unknown to them and in the creation of which, therefore, they have inwardly no share. A spiritual barrenness sets in and spreads, a chilling uniformity without height or depth. And bitterness awakes against the life vouchsafed to the *gifted* ones, the born *creators*. Men will no longer see, nor understand, that leaders' work is the *harder* work, and that their own life *depends* on its success; they merely sense that this work is making its doers happy, tuning and enriching the soul, and that is why they hate them.

TWELVE

IN reality, however, it is out of the power either of heads or of hands to alter in any way the destiny of machine-technics, for this has developed out of inward spiritual necessities and is now correspondingly maturing towards its fulfilment and end. To-day we stand on the summit, at the point when the fifth act is beginning. The last decisions are taking place, the tragedy is closing.

Every high Culture *is* a tragedy. The history of mankind *as a whole* is tragic. But the sacrilege and the catastrophe of the Faustian are greater than all others, greater than anything Æschylus or Shakespeare ever imagined. The creature is rising up against its creator. As once the microcosm Man against Nature, so now the microcosm Machine is revolting against Nordic Man. The lord of the World is becoming the slave of the Machine, which is forcing him — forcing us all, whether we are aware of it or not — to follow its course.

The victor, crashed, is dragged to death by the team.

At the commencement of the twentieth century the aspect of the "world" on this small planet is somewhat of this sort. A group of nations of Nordic blood under the leadership of British, Germans, French, and Americans commands the situation. Their political power depends on their *wealth*, and their wealth consists in their *industrial* strength. But this in turn is bound up with the existence of coal. The Germanic peoples, in particular, are secured by what is almost a monopoly of the known coal-fields, and this has led them to a multiplication of their populations that is without parallel in all history. On the ridges of the coal, and at the focal points of the lines of communication radiating therefrom, is collected a human mass of monstrous size, bred by machine-technics, working for it, and living on it. To the other peoples — whether in the form of colonies or of nominally independent states — is assigned the rôle of providing the raw material and receiving the products. This division of function is secured by armies and navies, the upkeep of which presupposes industrial wealth, and which have been fashioned by so thorough a technique that they, too, work by the pressing of a button. Once again the deep relationship, almost identity, of politics, war, and economics discloses itself. The *degree* of military power is dependent on the *intensity* of industry. Countries industrially poor are poor all round; they, therefore, cannot support an army or wage a war; therefore they are politically impotent; and, therefore, the workers in them, leaders and led alike, are pawns in the economic policy of their opponents.

In comparison with the masses of executive hands — who are the only part of the picture that discontent will look upon — the *increasing value* of the leadership-work of the few creative heads (undertakers, organizers, discoverers, engineers) is no longer comprehended and valued;[1] in so far as it is so at all, practical America rates it highest, and Germany, "the land of poets and thinkers," lowest. The imbecile phrase "The wheels would all be standing still, Did thy mighty arm so will" beclouds the minds of chatterers and scribblers. That even a sheep could bring about, if it were to fall into the machinery. But to invent these wheels and set them working so as to provide that "strong arm" with its living, that is something which only a few *born* thereto can achieve.

---

[1] *Decline of the West*, English edition, Vol. II, pp. 504 et seq.

These uncomprehended and hated leaders, the "pack" of the strong personalities, have a different psychology from this. They have not lost the old triumph-feeling of the beast of prey as it holds the quivering victim in its claws, the feeling of Columbus when he saw land on the horizon, the feeling of Moltke at Sedan as he watched the circle of his batteries completing itself down by Illy and sealing the victory. Such moments, such peaks of human experience, the shipbuilder, too, enjoys when a huge liner slides down the ways, and the inventor when his machine is run up and found to "go splendidly," or when his first Zeppelin leaves the ground.

But it is of the tragedy of the time that this unfettered human thought can no longer grasp its own consequences. Technics has become as esoteric as the higher mathematics which it uses, while physical theory has refined its intellectual abstractions from phenomena to such a pitch that (without clearly perceiving the fact) it has reached the pure foundations of human knowing.[1] *The mechanization of the world* has entered on a phase of highly dangerous over-tension. The picture of the earth, with its plants, animals, and men, has altered. In

---
[1] *Decline of the West*, English edition, Vol. I, pp. 420 et seq.

a few decades most of the great forests have gone, to be turned into news-print, and climatic changes have been thereby set afoot which imperil the land-economy of whole populations. Innumerable animal species have been extinguished, or nearly so, like the bison; whole races of humanity have been brought almost to vanishing-point, like the North American Indian and the Australian.

All things organic are dying in the grip of organization. An artificial world is permeating and poisoning the natural. The Civilization itself has become a machine that does, or tries to do, everything in mechanical fashion. We think only in horse-power now; we cannot look at a waterfall without mentally turning it into electric power; we cannot survey a countryside full of pasturing cattle without thinking of its exploitation as a source of meat-supply; we cannot look at the beautiful old handwork of an unspoilt primitive people without wishing to replace it by a modern technical process. Our technical thinking *must* have its actualization, sensible or senseless. The luxury of the machine is the consequence of a necessity of thought. In last analysis, the machine is a *symbol*, like its secret ideal, perpetual motion — a spiritual and intellectual, but no vital necessity.

It is beginning to contradict even economic practice in many ways. Already their divorce is being foreshadowed everywhere. The machine, by its multiplication and its refinement, is in the end defeating its own purpose. In the great cities the motor-car has by its numbers destroyed its own value, and one gets on quicker on foot. In Argentine, Java, and elsewhere the simple horse-plough of the small cultivator has shown itself economically superior to the big motor implement, and is driving the latter out. Already in many tropical regions the black or brown man with his primitive ways of working is a dangerous competitor to the modern plantation-technique of the white. And the white worker in old Europe and North America is becoming uneasily inquisitive about his work.

It is, of course, nonsense to talk, as it was fashionable to do in the nineteenth century, of the imminent exhaustion of the coal-fields within a few centuries and of the consequences thereof — here, too, the materialistic age could not but think materially. Quite apart from the actual saving of coal by the substitution of petroleum and water-power, technical thought would manage ere long to discover and open up still other and quite different sources of power. It is not worth while thinking ahead so far in time. For the West European-American technics *will itself have ended* by then. No stupid trifle like the absence of material would be able to hold up this gigantic evolution. So long as the thought working inside it is on its heights, so long will it be able without fail to produce the means for its purposes.

But *how long* will it stay on these heights? Even on the present scale our technical processes and installations, if they are to be maintained, require, let us say a hundred thousand outstanding brains, as organizers and discoverers and engineers. These must be strong — nay, even creative — talents, enthusiasts for their work, and formed for it by a steeling of years' duration at great expense. Actually, it is just this calling that has for the last fifty years irresistibly attracted the strongest and ablest of the white youth. Even the children play with technical things. In the urban classes and families, whose sons chiefly come into consideration in this connexion, there was already a tradition of comfort and culture, so that the normal preconditions were already provided for that mature and autumnal product, technical intellectuality.

But all this is changing in the last decades, in all the countries where large-scale industry is of old standing. The Faustian thought begins to be sick of machines. A weariness is spreading, a sort of pacifism of the battle with Nature. Men are returning to forms of life simpler and nearer to Nature; they are spending their time in sport instead of technical experiments. The great cities are becoming hateful to them, and they would fain get away from the pressure of soulless facts and the clear cold atmosphere of technical organization. And it is precisely the strong and creative talents that are turning away from practical problems and sciences and towards pure speculation. Occultism and Spiritualism, Hindu philosophies, metaphysical inquisitiveness under Christian or pagan colouring, all of which were despised in the Darwinian period, are coming up again. It is the spirit of Rome in the Age of Augustus. Out of satiety of life, men take refuge from civilization in the more primitive parts of the earth, in vagabondage, in suicide. *The flight of the born leader from the Machine is beginning.* Every big entrepreneur has occasion to observe a falling-off in the intellectual qualities of his recruits. But the grand technical development of the nineteenth century had been possible only because the intellectual level was constantly becoming higher. Even a stationary condition, short of an actual falling-off, is dangerous and points to an ending, however numerous and however well schooled may be the hands ready for work.

And how is it with them? The tension between work of leadership and work of execution has reached the level of a catastrophe. The importance of the former, the economic value of every real personality in it, has become so great that it is invisible and incomprehensible to the majority of the underlings. In the latter, the work of the hands, the individual is now *entirely* without significance. Only numbers matter. In the consciousness of this unalterable state of things, aggravated, poisoned, and financially exploited by egoistic orators and journalists, men are so forlorn that it is mere human nature to revolt against the rôle for which the machine (not, as they imagine, its possessors) earmarks most of them. There is beginning, in numberless forms — from sabotage, by way of strike, to suicide — *the mutiny of the Hands against their destiny*, against the machine, against the organized life, against anything and everything. The organization of work, as it has existed for thousands of years, based on the idea of " collective doing "[1] and the consequent division of

---

[1] See section 8 above.

labour between leaders and led, heads and hands, is being disintegrated from below. But " mass " is no more than a negation (specifically, a negation of the concept of organization) and not something viable in itself. An army without officers is only a superfluous and forlorn herd of men.[1] A chaos of brickbats and scrap-iron is a building no more. This mutiny, world-wide, threatens to put an end to the *possibility* of technical economic work. The leaders may take to flight, but the led, become superfluous, are lost. Their numbers are their death.

The third and most serious symptom of the collapse that is beginning lies, however, in what I may call *treason to technics*. What I am referring to is known to everyone, but it has never been envisaged in its entirety, and consequently its fateful significance has never disclosed itself. The immense superiority that Western Europe and North America enjoyed in the second half of the nineteenth century, in power of every kind — economic and political, military and financial — was based on an uncontested *monopoly* of industry. Great industries were only possible in connexion with the coal-fields

---

[1] What the Soviet régime has been attempting for the last fifteen years has been nothing but the restoration, under new names, of the political, military, and economic organization that it destroyed.

of these Northern countries. The rôle of the rest of the world was to absorb the product, and colonial policy was always, for practical purposes, directed to the opening-up of new markets and new sources of raw material, not to the development of new areas of production. There was coal elsewhere, of course, but only the white engineers would have known how to get at it. We were in sole possession, not of the material, but of the methods and the trained intellects required for its utilization. It is this that constitutes the basis of the luxurious living of the white worker — *whose income, in comparison with that of the " native," [1] is princely* — a circumstance that Marxism has turned to dishonest account, to its own ruin. It is being revenged on us today, for from now on, evolution is going to be complicated by the problem of unemployment. The high level of wages of the white worker, which is today a peril to his very *life*, rests upon the monopoly that the leaders of industry have created about him.[2]

And then, at the close of last century, the blind

---

[1] Including in this term the inhabitants of Russia and parts of southern and south-eastern Europe.

[2] Without going further afield, the tension that exists on the matter of wages between the land-worker and the metal-worker is evidence of this.

will-to-power began to make its decisive mistakes. Instead of keeping strictly to itself the technical knowledge that constituted their greatest asset, the "white" peoples complacently offered it to all the world, in every Hochschule, verbally and on paper, and the astonished homage of Indians and Japanese delighted them. The famous "dissemination of industry" set in, motivated by the idea of getting bigger profits by bringing production into the marketing area. And so, in place of the export of finished products exclusively, they began an export of secrets, processes, methods, engineers, and organizers. Even the inventors emigrate, for Socialism, which could if it liked harness them in its team, expels them instead. And so presently the "natives" saw into our secrets, understood them, and used them to the full. Within thirty years the Japanese became technicians of the first rank, and in their war against Russia they revealed a technical superiority from which their teachers were able to learn many lessons. Today more or less everywhere — in the Far East, India, South America, South Africa — industrial regions are in being, or coming into being, which, owing to their low scales of wages, will face us with a deadly competition. The unassailable privileges of the white races have been thrown away, squandered, betrayed. The others have caught up with their instructors. Possibly — with their combination of "native" cunning and the over-ripe intelligence of their ancient civilizations — they have surpassed them. Where there is coal, or oil, or water-power, there a new weapon can be forged against the heart of the Faustian Civilization. The exploited world is beginning to take its revenge on its lords. The innumerable hands of the coloured races — at least as clever, and far less exigent — will shatter the economic organization of the whites at its foundations. The *accustomed* luxury of the white workman, in comparison with the coolie, will be his doom. The labour of the white is *itself* coming to be unwanted. The huge masses of men centred in the Northern coal areas, the great industrial works, the capital invested in them, whole cities and districts, are faced with the probability of going under in the competition. The centre of gravity of production is steadily shifting away from them, especially since even the respect of the coloured races for the white has been ended by the World War. *This* is the real and final basis of the unemployment that prevails in the white countries. It is no mere crisis, but the *beginning of a catastrophe.*

For these "coloured" peoples (including, in this context, the Russians) the Faustian technics are in no wise an inward necessity. It is only Faustian man that thinks, feels, and *lives* in this form. To him it is a *spiritual* need, not on account of its economic consequences, but on account of its victories — "*navigare necesse est, vivere non est necesse.*" For the coloured races, on the contrary, it is but a weapon in their fight against the Faustian civilization, a weapon like a tree from the woods that one uses as house-timber, but discards as soon as it has served its purpose. This machine-technics will end with the Faustian civilization and one day will lie in fragments, *forgotten* — our railways and steamships as dead as the Roman roads and the Chinese wall, our giant cities and skyscrapers in ruins like old Memphis and Babylon. The history of this technics is fast drawing to its inevitable close. It will be eaten up from within, like the grand forms of any and every Culture. When, and in what fashion, we know not.

Faced as we are with this destiny, there is only one world-outlook that is worthy of us, that which has already been mentioned as the Choice of Achilles — better a short life, full of deeds and glory, than a long life without content. Already the danger is so great, for every individual, every class, every people, that to cherish any illusion whatever is deplorable. Time does not suffer itself to be halted; there is no question of prudent retreat or wise renunciation. Only dreamers believe that there is a way out. Optimism is *cowardice*.

We are born into this time and must bravely follow the path to the destined end. There is no other way. Our duty is to hold on to the lost position, without hope, without rescue, like that Roman soldier whose bones were found in front of a door in Pompeii, who, during the eruption of Vesuvius, died at his post because they forgot to relieve him. That is greatness. That is what it means to be a thoroughbred. The honourable end is the one thing that can *not* be taken from a man.

CPSIA information can be obtained
at www.ICGtesting.com
Printed in the USA
LVHW041259101220
673819LV00020B/449